Extreme Animals

Nature's Nastiest Biters

Frankie Stout

PowerKiDS press.

New York

MAI 8113013

For Nicholas Anthony Lazarus, a wonderful nephew

Published in 2008 by The Rosen Publishing Group, Inc.
29 East 21st Street, New York, NY 10010

First Edition

Editor: Jennifer Way
Book Design: Greg Tucker
Photo Researcher: Nicole Pristash

Photo Credits: Cover, pp. 7, 9, 13, 15, 16 (top left), 21 Shutterstock.com; p. 5 © www.iStockphoto.com/WinterWitch; p. 11 © Klein/Peter Arnold, Inc.; p. 16–17 (main) © Klaus Jost/SeaPics.com; p. 19 © Pete Oxford/Getty Images.

Library of Congress Cataloging-in-Publication Data

Stout, Frankie.
  Nature's nastiest biters / Frankie Stout. — 1st ed.
      p. cm. — (Extreme animals)
  Includes index.
  ISBN 978-1-4042-4157-2 (lib. bdg.)
  1. Animal attacks—Juvenile literature. 2. Bites and stings—Juvenile literature. I. Title.
  QL100.5.S86 2008
  591.6'5—dc22
                                        2007026870

Manufactured in the United States of America

Contents

Extreme Biters

Some animals are known for their powerful or nasty bite. The world's most **extreme** biters can use their powerful **jaws** and teeth to catch **prey** or to keep **predators** away. Sometimes these animals can even use their bite to **poison** their enemies.

These extreme biters come in all sizes and shapes and live in all kinds of **habitats**. Sharks, hyenas, crocodiles, snakes, and many other types of animals all use their bite to **survive** in water and on land. From Komodo dragons to Tasmanian devils, animals in the natural world will wow you with their extreme bites.

The Komodo dragon is the world's biggest lizard. Each dragon has around 60 inch-long (2.5 cm) teeth. These teeth fall out and new ones grow back throughout the Komodo dragon's life.

Built for Biting

Extreme biters use their teeth and jaws to survive. Their bite helps them catch food or scare away other animals.

Many extreme biters are carnivores, or meat eaters. Carnivores have long, sharp teeth that can cut, rip, and tear the body of animals they catch. Some carnivores, like crocodiles, have many pointed teeth in large jaws that catch and cut through big prey quickly. Crocodiles are some of the nastiest biters around. They have more than 60 teeth!

The American crocodile lives in wetlands on the southern tip of Florida.

Biters on the Hunt

Hyenas are carnivores that hunt at night. Hyenas' strong jaws and teeth help them eat every part of the animals they catch. Thanks to their powerful bite, hyenas can break the bones and even the horns of their prey! Because a hyena may go days without eating, it has to eat every last part of its catch.

Adult hyenas can put 800 pounds per square inch (56 kg/sq cm) of force into their bite. They have some of the strongest jaws of any animal.

Because of their nasty bite, hyenas have few enemies. Lions and tigers are just about the only animals hyenas try to stay away from!

A Night Biter

Tasmanian devils live on the island of Tasmania, which is near Australia. They are marsupials. Marsupials are animals whose babies are born very small and stay in a pouch on their mother's stomach until they are bigger.

Tasmanian devils have a wild cry that makes them sound scary. They also have big jaws that open wide and make them look scary. Tasmanian devils like to look for food at night. However, Tasmanian devils do not always feed on animals that they have killed. Many of their meals are animals, such as wallabies and sheep, that are already dead. This is called scavenging.

Tasmanian devils use their strong jaws and big teeth to eat all of a dead animal, even the bones and fur!

The Hippopotamus's Big Bite

Hippopotamuses, or hippos, do not bite to catch prey but to keep predators away. Hippos live in Africa, in rivers, lakes, and other wet places. They stay in water and feed on water plants during the day. At night, hippos leave the water and feed on land. Hippos share their habitats with big carnivores like lions and crocodiles.

Crocodiles trying to catch baby hippos for a meal must get past angry hippo mothers. Adult hippos have long front teeth growing outside their mouth. These teeth can grow to be almost 2 feet (61 cm) long. A hippo mother trying to save her young can bite through the body of a crocodile!

These hippos are fighting. Male, or boy, hippos fight to stake out their territory. They will use their sharp teeth to try to cut and bite deeply into the other hippo.

The Biters of the Deep

One of the world's nastiest biters is the great white shark. Great white sharks have rows of triangular pointed teeth. Sometimes these teeth break off. When a tooth breaks, another tooth from the row behind it moves up to take its place in the shark's mouth!

Great white sharks like to eat other fish, seals, dolphins, and even turtles. A great white shark on the hunt will first take a huge bite that kills its prey. The shark then backs off and waits for the animal to die before it eats its prey.

Great white sharks can grow to be 20 feet (6 m) long and weigh 1,900 pounds (862 kg).

Great White Shark

The teeth of great white sharks have sharp edges, like a saw.

WoW!!

A great white shark can bite off up to 30 pounds (14 kg) of meat at a time!

Big Fish!

An adult great white can weigh up to 5,000 pounds (2,268 kg).

Extreme Facts

1. Great white sharks live in all the world's oceans.

2. Great white sharks are often found near coasts.

3. Great white sharks often attack their prey from below.

4. Great white sharks are the largest predators of fish.

Snakes and Bites

Poisonous snakes use their hollow, sharp teeth to poison other animals. They do this to kill their prey and to attack animals when they feel they are in danger.

A snake on the hunt will strike quickly and then try to move away from its prey. The prey soon dies from poisoning.

Although sharks are much bigger than snakes, many more snakes than sharks bite people every year. Some snakes' bites are more poisonous than others'. There are poisonous snakebites, like the black mamba's, that are almost always deadly to humans.

This eyelash viper is eating a mouse whole. The viper killed the mouse with the deadly poison that flowed through the viper's teeth as it bit the mouse.

Biting for a Living

All kinds of animals use their bite to catch food or to **defend** themselves. Special teeth and strong jaws can help an animal survive in its habitat. Without their big teeth, Tasmanian devils and crocodiles could not eat, and hippo mothers could not defend their babies.

Though they can look scary, with their sharp teeth and big jaws, extreme biters can be helpful, too. Some keep their habitats clean by eating dead animals. Biters that eat meat also keep a habitat from having too many animals **competing** for food. Extreme biters are some of nature's most interesting animals!

The mighty lion is known for its powerful jaws and killer hunting skills. Because of that, these big cats have few animals that will try to hunt them!

Quick Bites

Sharks can grow more than 20,000 teeth in their lifetime!

Dolphins can have up to 200 teeth.

Most **people** have 32 teeth.

Crocodiles have about 60 teeth.

Elephant tusks can grow to be 5 feet (1.5 m) long.

Rabbits' teeth never stop growing but are kept short by chewing on lots of hard foods.

Glossary

competing (kum-PEET-ing) Trying to get something that another animal wants.

defend (dih-FEND) To guard from harm.

extreme (ik-STREEM) Going past the expected or common.

habitats (HA-beh-tats) The kinds of land where animals or plants naturally live.

jaws (JAHZ) Bones in the top and bottom of the mouth.

poison (POY-zun) To cause pain or death with matter made by an animal's body.

predators (PREH-duh-terz) Animals that kill other animals for food.

prey (PRAY) An animal that is hunted by another animal for food.

survive (sur-VYV) To stay living.

tusks (TUSKS) Long, large pointed teeth that come out of the mouth of some animals.

Index

C

crocodile(s), 4, 6, 12, 20, 22

F

food(s), 10, 20, 22

H

habitat(s), 4, 12, 20
hyenas, 4, 8

J

jaws, 4, 6, 8, 10, 20

K

Komodo dragons, 4

L

land, 4

M

meat, 16, 20

P

predators, 4, 12, 17
prey, 4, 6, 8, 12, 14, 17–18

S

sharks, 4, 14, 16–18, 22
snake(s), 4, 18

T

Tasmanian devils, 4, 10, 20
tusks, 22
types, 4

W

water, 4, 12
world, 4

Web Sites

Due to the changing nature of Internet links, PowerKids Press has developed an online list of Web sites related to the subject of this book. This site is updated regularly. Please use this link to access the list:
www.powerkidslinks.com/exan/bite/